LET'S CODE!

CODING LOOPS

BY GEORGE ANTHONY KULZ

CONTENT CONSULTANT
DAVID A. LASH
ASSOCIATE PROFESSOR OF COMPUTER SCIENCE
AURORA UNIVERSITY

Kids Core
An Imprint of Abdo Publishing
abdobooks.com

abdobooks.com

Published by Abdo Publishing, a division of ABDO, PO Box 398166, Minneapolis, Minnesota 55439.

Printed in the United States of America, North Mankato, Minnesota.
102023
012024

THIS BOOK CONTAINS RECYCLED MATERIALS

Cover Photo: Africa Studio/Shutterstock Images
Interior Photos: Dragon Images/Shutterstock Images, 4–5; Shutterstock Images, 7, 22, 25 (aliens), 29 (aliens); LightField Studios/Shutterstock Images, 9, 28 (top); Jeppe Gustafsson/Shutterstock Images, 10; Ground Picture/Shutterstock Images, 12–13; Robyn Beck/AFP/Getty Images, 15; Sergey Novikov/Shutterstock Images, 16, 28 (bottom); Ashan Randika/Shutterstock Images, 18–19; Jacob Lund/Shutterstock Images, 20; Kevork Djansezian/Getty Images News/Getty Images, 23; Red Line Editorial, 25 (flowcharts), 29 (flowcharts); Rijasolo/AFP/Getty Images, 26

Editor: Katharine Hale
Series Designer: Katharine Hale

Library of Congress Control Number: 2023939662

Publisher's Cataloging-in-Publication Data

Names: Kulz, George Anthony, author.
Title: Coding loops / by George Anthony Kulz
Description: Minneapolis, Minnesota: Abdo Publishing, 2024 | Series: Let's code! | Includes online resources and index.
Identifiers: ISBN 9781098292775 (lib. bdg.) | ISBN 9798384910718 (ebook)
Subjects: LCSH: Coding theory--Juvenile literature. | Computer programming--Juvenile literature. | Computers and children--Juvenile literature. | Loops (Computer science)--Juvenile literature. | Programming loops--Juvenile literature.
Classification: DDC 005.1--dc23

CONTENTS

Some people concentrate better when listening to music.

CHAPTER 1

ON REPEAT

Jeremy walks into his bedroom and drops his backpack near his desk. He has a lot of homework to do tonight. Jeremy likes to listen to music while doing homework. It helps him focus. Jeremy opens a phone app.

The app contains a playlist of all his favorite songs. He taps “play” on the playlist. All the songs on the list will play. Jeremy will not have to hit “next.” The app takes care of playing music for him.

Inside Jeremy’s phone is a small computer. His music app runs on that small computer. The app plays his songs until they are done. The music app uses something called a loop to do this.

What Is a Loop?

Jeremy’s music app is a program. Programs are written in computer code. They give instructions to a computer. Programmers are people who use code to write programs or **software**.

Programmers write code in programming languages.

Loops are an important part of many programs. A loop is a block of steps that are repeated. Suppose a set of steps needs to happen 100 times. Writing out the steps would take a long time. Programmers use loops instead. A programmer writes the set of

Loops and Lists

Loops can be used to create lists. Someone might want to create a list of the even numbers between zero and 100. Rather than typing them all out, a loop can be used. Starting at zero, the program can add two each time through the loop. It creates a list containing zero, two, four, six, and so on. The loop stops when it reaches 100.

People can easily repeat steps for an activity such as jumping rope. But computers need to be told how to do each step.

steps once. Then they use a loop. The loop tells the computer to repeat the steps 100 times.

All loops have a condition. A condition is a statement that is either true or false. If a loop's condition is true, the loop's steps will repeat.

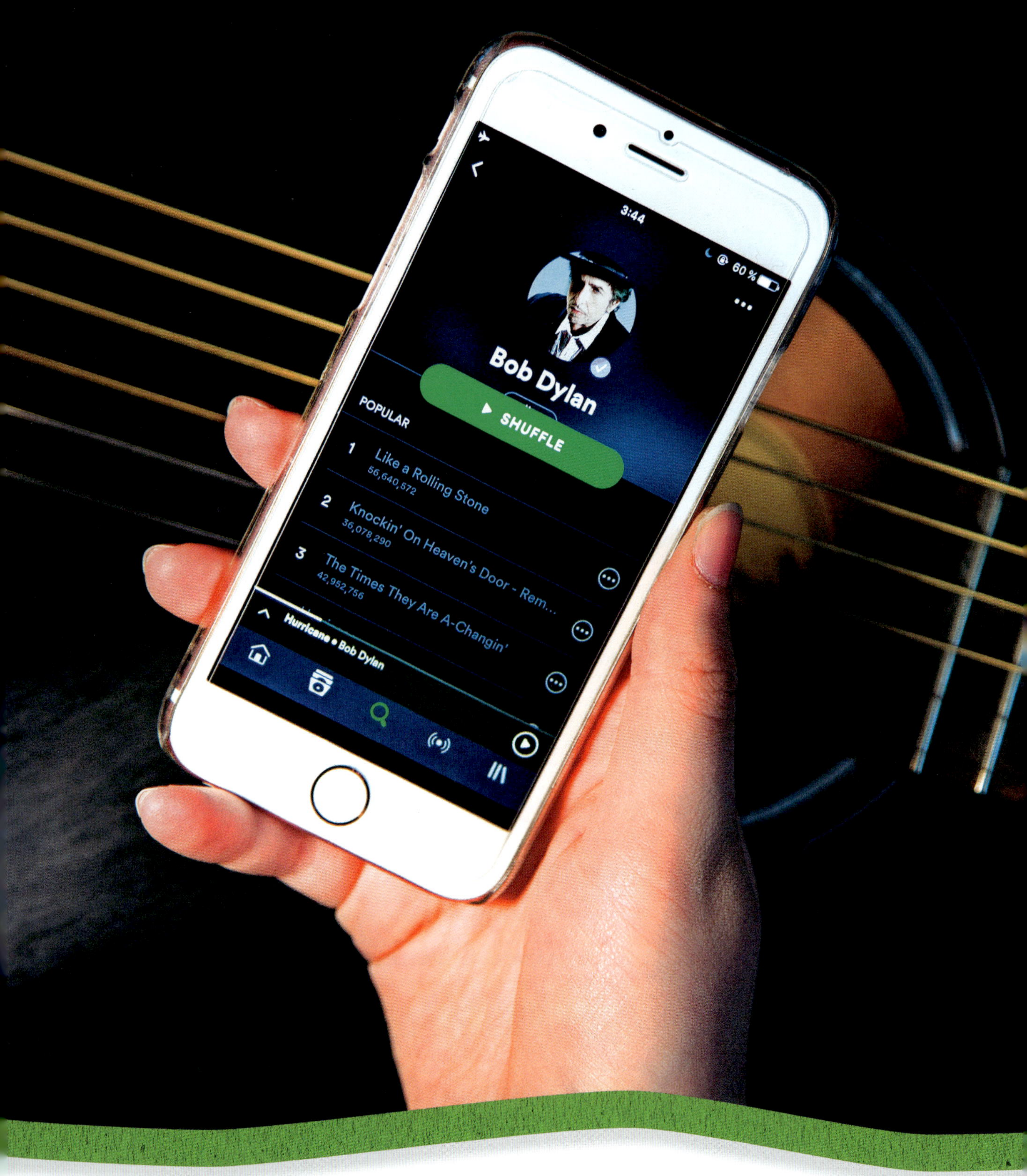

Playlists let people put songs together that will play one after the other.

Once the condition becomes false, the loop will stop.

Jeremy's phone plays his music in a loop. The app will play the first song. When the song is done, the app checks a condition. The condition is, "Are there more songs in the list?" If this is true, the app plays the next song. This continues until it reaches the last song. The condition will then become false, and the process will stop.

Explore Online

Visit the website below. Does it give any new information about loops that wasn't in Chapter One?

Sequences, Selections, and Loops

abdocorelibrary.com/coding-loops

Planning out a program can help programmers decide whether they should use loops.

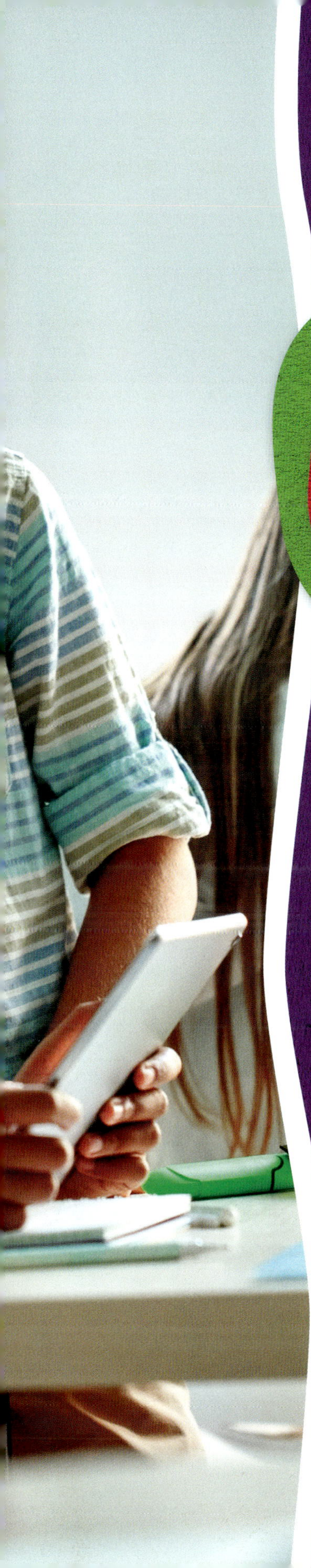

CHAPTER 2

STARTING A LOOP

Loops are powerful tools for writing code. But they are only used to solve certain kinds of problems. Programmers need to know when and how to use them. Planning ahead can help a programmer know if loops would be useful in their program.

A programmer can write out all the steps to the program. If any steps repeat, the programmer can use a loop. This will save them from writing the same code many times in a program.

Using Loops

Programmers use loops to repeat steps in a program. Sometimes the same exact

Loops inside Loops

Loops can be placed inside other loops. The inside loop is called a **nested** loop. A program that counts hours and minutes is an example of a nested loop. The hour number increases once per hour. The minute number increases once per minute. Both loops are running at the same time.

Programmers can use loops to make a character do the same move over and over again.

step repeats. One example would be a character doing a dance move. A programmer could loop the same dance move until they want the character to stop.

But sometimes the repeated step needs to change. A programmer might work on a hide-and-seek program. A character needs to count to 10. Then the character says, “Ready or not, here I come!” The programmer could write, “Say 1,” “Say 2,” all the way up to 10.

Programs can count using loops.

But that would take a long time. Instead, the programmer can write a loop. They tell the loop to start at one. Then they tell the character to say the number plus one. Each time the character counts, the number goes up.

The steps still repeat. But there is a small change every time.

Once a programmer knows what needs to loop, they need to decide what kind of loop to use. The kind of loop determines how the loop will end. Most loops need a clear stopping point. Otherwise, the loops could run forever. This can cause a program to **crash**.

Further Evidence

Watch the video at the website below. Does it give any new evidence to support Chapter Two?

Intro to *While* Loops

abdocorelibrary.com/coding-loops

Loops that don't end can cause programs to crash.

CHAPTER 3

ENDING A LOOP

There are two basic types of loops. These are *for* loops and *while* loops. Both loop types can run the same steps. But they end differently.

Sometimes programmers know exactly how many times their loop needs to run.

Some loops run a specific number of times. Others run until a certain condition is met.

When this happens, they use a *for* loop. A *for* loop runs for a specific number of **iterations**. An iteration is one time through the loop. When the last iteration finishes, the loop ends. The loop is checking a simple condition. The condition is whether or not the number of iterations is complete.

Sometimes programmers do not know when their loop needs to stop. This situation requires a *while* loop. *While* loops use unique conditions. A programmer defines the condition before starting the loop. The steps in a *while* loop will repeat until this condition is met. When it is, the loop ends.

The Infinite Loop

Sometimes the test condition in a loop is never met. This is called an **infinite** loop. Most times this is an error in the code. But some infinite loops are good. Gaming consoles run games in an infinite loop until the player quits.

Video games can use both *for* and *while* loops.

For vs. *While*

A programmer could write a game where a player needs to defeat enemies. The programmer could write a loop. Every five seconds, the game would **spawn** an enemy.

Many video games involve defeating enemies.

The programmer would need to decide how many times to run the loop. They could make a set number of enemies appear. The programmer would use a *for* loop to do this. For example, the game would spawn one enemy every five seconds. It would do this ten times. Once the game has created ten enemies, the loop will stop. If the player survived, they would move to the next level.

The programmer could also have the player level up based on points. The player could earn points for every enemy defeated. Some enemies could be worth more points than others. As soon as the player reaches 100 points, they would move to the next level.

The level would not have a set number of enemies this way. The programmer can use a *while* loop instead. While the player has fewer than 100 points, the game will spawn one enemy every five seconds. The game is constantly checking the player's score. When the player reaches 100 points, the condition is met. They move immediately to the next level. *For* and *while* loops have different purposes. But they both would work to make the game fun.

For and *While* Loops

```
For (10 times)
Loop
   Spawn enemy
   Wait 5 seconds
```

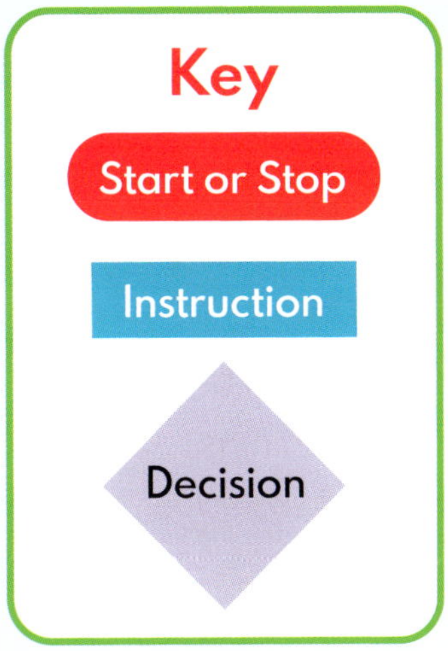


```
While (score < 100)
Loop
   Spawn enemy
   Wait 5 seconds
```

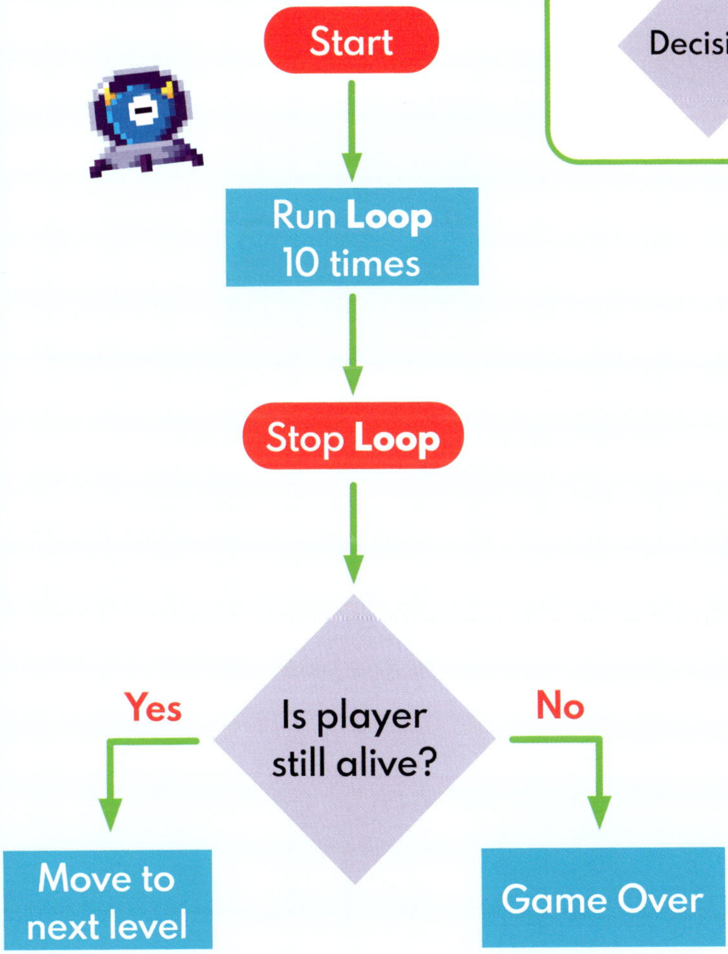

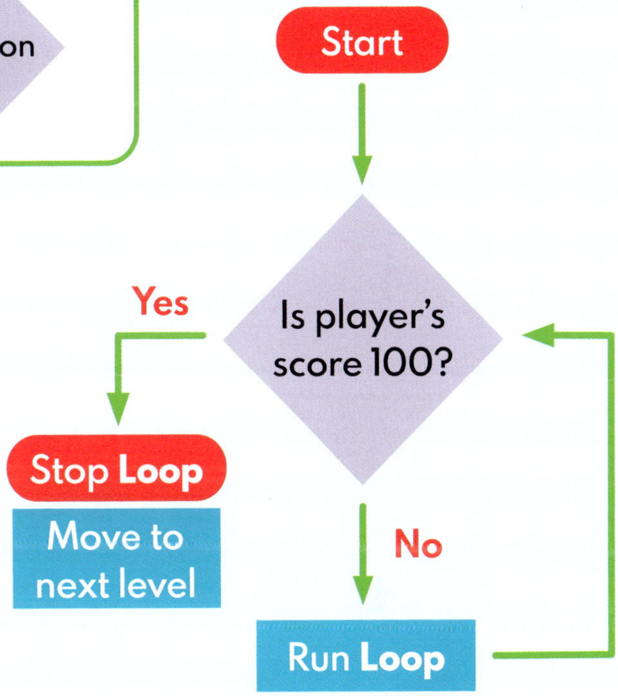

For and *while* loops serve different purposes in programming.

Loops help programmers save time while writing code.

Loops are powerful tools for writing programs. Writing a loop takes fewer lines of code than writing all the repeated steps. Choosing the right loop will make coding faster and easier.

PRIMARY SOURCE

Jonathan Hsu writes about software development. He discussed how to decide what kind of loop to use:

> My universal rule for choosing between a *while* loop and *for* loop is that I use *while* loops when I do not know the number of iterations ahead of time and *for* loops when I do know.

Source: Jonathan Hsu. "How to Pick between a For Loop and While Loop." *Built In*, 25 Aug. 2022, builtin.com. Accessed 1 Apr. 2023.

Point of View

What is the author's point of view on this topic? What is your point of view? Write a short essay about how they are similar and different.

CODING DATA

A loop is a block of steps in a program that are repeated.

Loops can repeat the exact same steps over and over again. They can also repeat steps with slight changes, such as counting up.

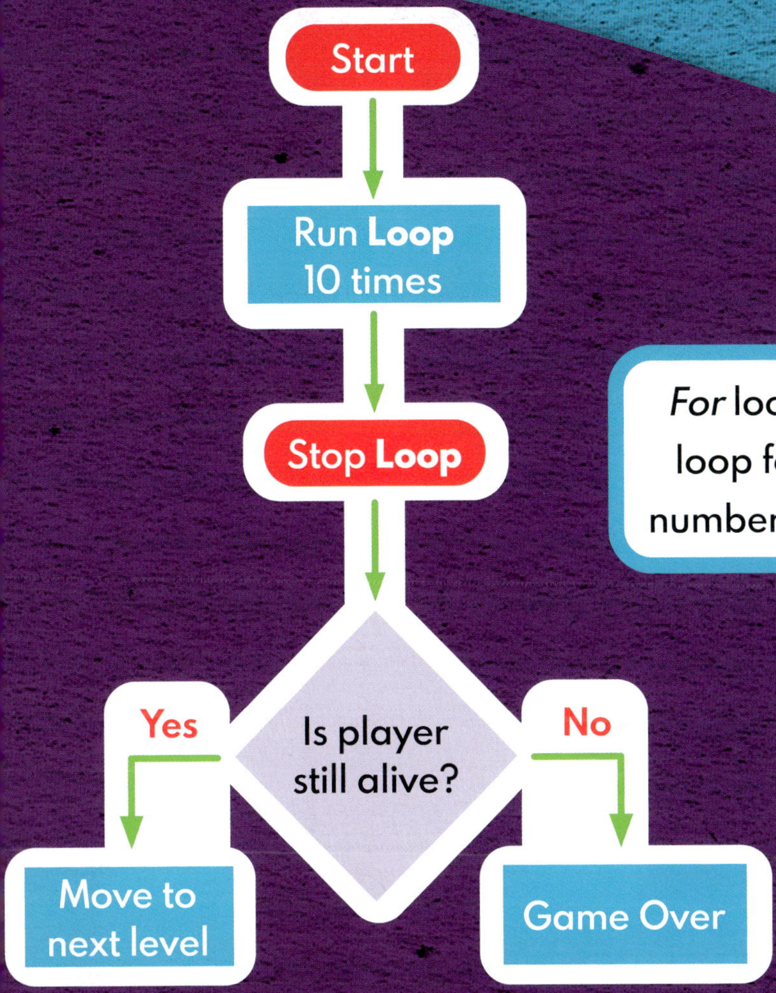

For loops repeat a loop for a specific number of iterations.

While loops repeat a loop until a unique condition is met.

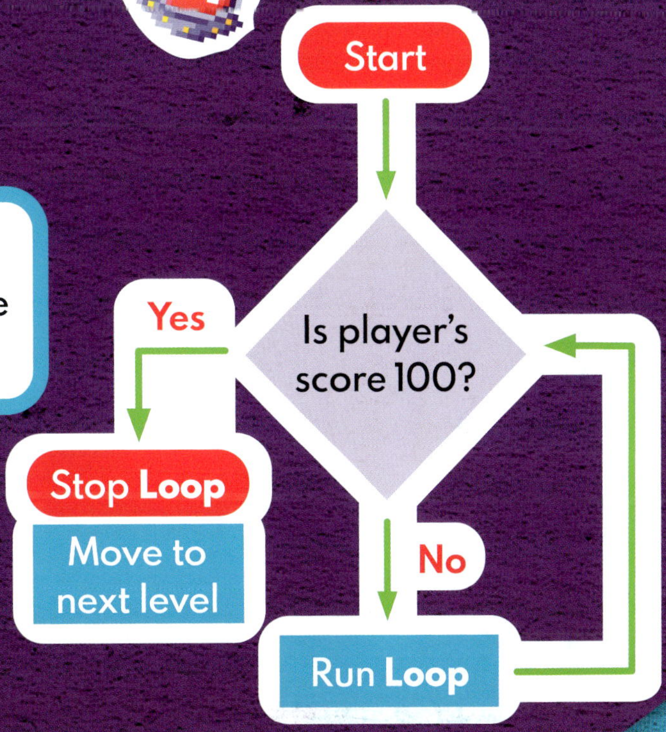

Glossary

crash
when a program suddenly stops working

infinite
unending, or extending forever

iteration
one time through a loop

nested
placed one inside another

software
all of the programs that give a computer instructions

spawn
to create

Online Resources

To learn more about loops, visit our free resource websites below.

Core Library CONNECTION
FREE! COMMON CORE MULTIMEDIA RESOURCES

Visit **abdocorelibrary.com** or scan this QR code for free Common Core resources for teachers and students, including vetted activities, multimedia, and booklinks, for deeper subject comprehension.

Booklinks NONFICTION NETWORK
FREE! ONLINE NONFICTION RESOURCES

Visit **abdobooklinks.com** or scan this QR code for free additional online weblinks for further learning. These links are routinely monitored and updated to provide the most current information available.

Learn More

Holmes, Kirsty. *Perfect Program!* Crabtree, 2019.

McKinney, Donna B. *Coding Conditionals.* Abdo, 2024.

Prottsman, Kiki. *How to Be a Coder.* DK, 2019.

Index

About the Author

George Anthony Kulz holds a master's degree in computer engineering. He is a member of the Society of Children's Book Writers and Illustrators and has taken courses at the Institute of Children's Literature and the Gotham Writers' Workshop. He writes for children and adults.